Jonathan Townsend

Identifying Important Habitat Features for Bat Conservation

Jonathan Townsend

Identifying Important Habitat Features for Bat Conservation

Using Acoustic Sampling and Geographic Information Systems

LAP LAMBERT Academic Publishing

Impressum / Imprint

Bibliografische Information der Deutschen Nationalbibliothek: Die Deutsche Nationalbibliothek verzeichnet diese Publikation in der Deutschen Nationalbibliografie; detaillierte bibliografische Daten sind im Internet über http://dnb.d-nb.de abrufbar.

Alle in diesem Buch genannten Marken und Produktnamen unterliegen warenzeichen-, marken- oder patentrechtlichem Schutz bzw. sind Warenzeichen oder eingetragene Warenzeichen der jeweiligen Inhaber. Die Wiedergabe von Marken, Produktnamen, Gebrauchsnamen, Handelsnamen, Warenbezeichnungen u.s.w. in diesem Werk berechtigt auch ohne besondere Kennzeichnung nicht zu der Annahme, dass solche Namen im Sinne der Warenzeichen- und Markenschutzgesetzgebung als frei zu betrachten wären und daher von jedermann benutzt werden dürften.

Bibliographic information published by the Deutsche Nationalbibliothek: The Deutsche Nationalbibliothek lists this publication in the Deutsche Nationalbibliografie; detailed bibliographic data are available in the Internet at http://dnb.d-nb.de.

Any brand names and product names mentioned in this book are subject to trademark, brand or patent protection and are trademarks or registered trademarks of their respective holders. The use of brand names, product names, common names, trade names, product descriptions etc. even without a particular marking in this work is in no way to be construed to mean that such names may be regarded as unrestricted in respect of trademark and brand protection legislation and could thus be used by anyone.

Coverbild / Cover image: www.ingimage.com

Verlag / Publisher:
LAP LAMBERT Academic Publishing
ist ein Imprint der / is a trademark of
OmniScriptum GmbH & Co. KG
Heinrich-Böcking-Str. 6-8, 66121 Saarbrücken, Deutschland / Germany
Email: info@lap-publishing.com

Herstellung: siehe letzte Seite /
Printed at: see last page
ISBN: 978-3-659-71920-2

Zugl. / Approved by: Fredonia, SUNY Fredonia, 2014

Copyright © 2015 OmniScriptum GmbH & Co. KG
Alle Rechte vorbehalten. / All rights reserved. Saarbrücken 2015

Identifying Important Habitat Features for Bat Conservation Using Acoustic Sampling and GIS

By

Jonathan Peter Townsend

Table of Contents

Abstract..3

Introduction..4

Methods..12
    Habitat..12
    Bat Surveys...14
    Species Identification..16
    Geo-spatial Analysis...18
    Statistical Analysis...20

Results...20

Discussion..30

Acknowledgements..35

Literature Cited...36

Appendix..38

List of Tables

Table 1. Summary of Habitat Type Definitions..13

Table 2. Typical Parameters for Identification of Bat Sonar Calls in NYS..18

Table 3. Bat calls identified from summer 2013 acoustic surveys and associated sample size for analysis...................23

Table 4. Significant Interactions Between Species and Habitat Type..24

Table 5. Significant Habitat to Habitat Interactions..27

List of Figures

Figure 1. Observed versus Expected Habitat Type Frequency as Compared With Bat Calling Activity......................22
"E" refers to the expected value of call activity based on frequency of habitat types in the environment, while "O" refers to the actual observed value of calling activity during the study.

Figure 2. Bat Calling Activity as Compared Across Species and Habitats. 25
All habitat types were compared with the 5 species that had sufficient sample size for discussion.

Figure 3. Bat Call Activity and Habitat Type. 26
The habitats that had significant interactions through loglinear analysis are compared with the species that had sample N > 50.

Figure 4. Histogram illustrating calling frequency with elevation. 29

Abstract

Bat populations worldwide have been under pressure for decades due to loss of habitat, roost disturbances and environmental toxins. Recently a fungus causing White Nose Syndrome has been infecting bat hibernacula in the United States, and to date has killed almost 6 million bats. In order to improve bat conservation efforts, habitat delineations and bio-acoustical sampling were conducted along two transects in Chautauqua County, NY from mid-May until the end of August, 2013. Surveys were vehicular, and driven between 29 - 32 kmph in order to match bats flying speed. They were conducted 30 min after sunset on nights where the temperature was > 13 °C. Twenty surveys were completed, and 1248 bats were identified to species. Loglinear analysis revealed a significant relationship between bat calling activity and forested habitats, specifically for big brown, silver haired, eastern red, and hoary bats. Wetland, stream, and residential habitats as well as elevation were also shown to have a significant relationship with calling activity. This study supports the hypothesis that bats forage in somewhat different habitats at the species level, and indicates the relatively strong importance of forested areas to bats. Additionally, the methodology for this study has the potential to gather rather large data sets in a short period of time, while collecting data on several species of bat at once.

Introduction

Bat species are extremely important to the ecosystems they inhabit. In tropical climates bats are responsible for propagating and spreading seeds from many fruit bearing trees, among other ecologically and economically important crops. More relevant to Chauatauqua County, NY is the impact of insect eating species. Insectivorous bat species not only reduce the potential for mosquito borne illness by consuming an extraordinary amount of pathogen spreading insects, they also reduce the damage done to the agricultural industry from crop destroying pests. For example, a colony of 150 big brown bats (*Eptesicus fuscus*) has been estimated to consume or suppress the following, quite impressive, number of agricultural pests in just one year: 600,000 cucumber beetles; 33,000,000 rootworms; 335,000 stink bugs; 194,000; scarab beetles; and 158,000 leafhoppers. The impact of this one species of bat affects crop harvests of cucumbers, corn, soybeans, cotton, potatoes, apples, as well as lawn and nursery industries by driving pest populations down (Agosta, 2002).

Chautauqua County is home to a large agricultural industry. Based on the 2007 Census of Agriculture, the county holds 1,658 farms, which comprises 95,448 hectares and in 2007 crop sales amounted to $57,810,000. The primary crop is grapes; in fact Chautauqua County is the number one producer of grapes in the state. Other crops include corn, potatoes, and grain, as well as various fruits, berries and nuts (Carlberg, 2009). Considering that grapes are the number one crop in the county, a review of the pests specific to that industry and how bats factor in is crucial. The most detrimental pest insect on vineyards is the grape berry moth, which is widespread and generally can be expected to damage grape yields on an annual basis. Other pest species include the steely beetle, plant bugs, grape leafhoppers, potato leafhopper, grape phylloxera, grape rootworm, Japanese beetle, European red mite, and grape mealybug (Loeb, 2012). Factor in the impact of just big brown bats (many of the species of insect consumed by this species are listed above), and it is quickly apparent that bat species have a beneficial impact on agriculture in the county.

Agricultural benefits aside, bat species have been long noted for their ecological services. In addition to acting as a natural check in the insect populations, bats often excrete guano on the landscape as they fly. Guano is high in nitrogen and considered to be an excellent fertilizer therefore, this nightly dispersal of nutrients is important in the larger ecological web (especially in tropical regions). Guano is also actively traded at market values, bringing a direct monetary value to their contribution (Kunz *et al.*, 2011). Recent studies have suggested that bats also have value as bio-indicators and indicators of ecological disturbance (Jones *et al.*, 2009 and Medellin *et al.*, 2000), further enhancing their importance in the local and global frameworks.

Despite their importance and uniqueness, bats have been persecuted throughout history as omens of evil, pests, and ironically enough given their consumption of disease spreading organisms; they themselves are often seen as a health concern. In the past several decades bat species have been decimated through environmental degradation. Habitat fragmentation, deforestation, and disease have very negatively affected bat populations world wide. Many species are extremely slow to reproduce, with females only producing one to two pups a year on average and these pups face high mortality. It is quickly apparent that populations decline can take bats many years to come back from.

Some of the major issues affecting bat populations currently are White Nose Syndrome (WNS), which is the outcome of a fungal infection caused by the species *Psuedogymnoascus destructans*, and to date has killed nearly 6 million bats in the northeastern US; the rabies virus, although it is not as prevalent in bats as many are led to believe (Klug *et al.*, 2011); and wind energy, which kills bats through direct collisions and barotrauma (Baerwald *et al.* Vol. 18 no. 16 R696), and affects mostly *Lasiurus* species over the *Myotis* species hard hit by WNS.

All of the bat species in New York State are insectivorous, echolocating species. There are three phases of a bat's echolocation including search, approach, and terminal phase. The search phase call is the most standard form of a bat's call and is the phase used in species identification. The approach phase begins to be emitted as a bat starts to receive information suggesting the availability of prey. Compared with search phase call, duration, inter-call interval, and frequency all decrease for approach phase calls. Terminal phase calls, or the "feeding buzz", have frequency ranges and durations that continue to decrease until the prey is caught (Harvey *et al.*, 2011). Depending on species and phase, an echolocation call can provide information ranging from distance to prey/obstruction, speed of approaching prey through use of doppler shift, and the size and shape of prey which allows specialists to select their favorite insect.

Acoustic survey research has been conducted on many different types of vocalizing animals in the past. Amphibians have been studied for decades using acoustic sampling. In a similar fashion to bat species, frogs and other amphibians have been the subject of communication and presence/absence experiments (Narins, 1995). Avian species present many of the same obstacles to scientists as bats as flying animals they can often be difficult to locate and record therefore automated acoustic sampling has also been a goal of avian biologists (Brandes, 2008). Also, insects are identifiable through acoustic sampling, and often have interactions with bat species based on ultrasonic triggers in their auditory systems, yet another facet to bat ecology (Mankin *et al.*, 2011). Cetaceans are mammal species that have echolocating abilities as well, and have previously been the subject of acoustic experiments and survey.

Recently bats have become a focal point for ultrasonic survey research, which is being used to determine species biodiversity as well as examine communication among species. Considering how complex social behavior and the associated ultrasonic communication can become in bats (Yovel *et al.*, 2009), it should come as no surprise that most bats have very species-specific call structures. The science is still in its early stages, yet most studies are indicating acoustic surveys as not only an appropriate method of species identification and populations and habitat studies, it also has the benefit of a less invasive approach that avoids stress on the study subjects. When factoring in WNS, this method is of huge import since one of the most difficult aspects of studying and combating the spread of WNS is preventing cave to cave transmission. Recording bat calls rather than individually capturing bats or entering caves is not only cheaper, it can often provide larger amounts of data in shorter lengths of time, and decreases the impact on the bats being surveyed.

There are disadvantages well as advantages to acoustic survey research for bats. As stated above, being able to study bats without netting them in mist nets or harp traps (which can stress and/or injure the bat), or disturbing hibernating species in cave systems, is a great step forward. However, there are a few issues to consider and attempts should be made to mitigate the impact of these issues in acoustic surveys. Stationary acoustic surveys are often confounded by the fact that it is impossible to differentiate between one bat circling the microphone and several bats passing through. This necessitates that the researcher conduct mist netting along with the acoustic study, negating many of the benefits just mentioned. The somewhat recent advent of vehicular acoustic studies has alleviated this issue. By attempting to match the bat's flying speed (driving 29 - 32 kmph) the vehicle quickly enters and exits the individual flying bat's range, and the researcher has greater assurance that each call recorded is a separate individual. Stationary acoustic surveys can also be confounded by the impact of the habitat where the recording is taking place. The degree of obstacles, or clutter, in the environment can impact bat call structure, perhaps making it more difficult to distinguish species because calls are more variable. Vehicular acoustic surveys afford a greater degree of standardization since the calls are recorded while bats are flying over roadways with less clutter.

Determining roosting locations for bats, while challenging as well, is an easier task than determining foraging locations due to the challenges of following or locating flying animals traveling from 0.5 - 10 km. to forage (Lacki *et al.*, 2007). There has been information gathered on foraging locations through mist netting during peak activity periods of foraging (after dusk and before dawn) as well as more recent radiotracking studies. These focus on a single or a small number of habitats where mist nets are set up, or focus on the foraging of a few individual bats. There are also limitations to these methods in that they require larger numbers of people to conduct. Monitoring of mist nets for several hours as well as the need for multiple teams if multiple locations are being surveyed at the same time in different locations is prohibitive. Radiotracking also requires multiple individuals to conduct and would need multiple teams to track the flight of multiple bats. Vehicular acoustic surveying allows for a focus on a larger scale (regional/population level) including activity of many bats and species as well as the benefit of the need for only one individual to gather these data.

There are 9 species of bat in New York State (Whitaker *et al.*, 1998), that are often separated into two categories including migratory; tree roosting species and the hibernating cave dwelling species (Lacki *et al.*, 2007). Generally, the migratory species are the silver haired (*Lasionycteris noctivagans*), Hoary (*Lasiurus cinereus*), and the eastern red (*Lasiurus borealis*). Silver haired bats have been found to forage over woodland ponds and streams (through mist netting) (Harvey *et al.*, 2011). Hoary and eastern red bats have been shown to consume primarily moths (through fecal analysis) perhaps indicating a foraging habitat of predominately forests. However, hoary bats, (often considered a "moth specialist") have been shown to shift diet based on the structural complexity of the foraging habitat and eastern red bats will also consume beetles opportunistically (Carter *et al.*, 2003).

Big brown bats (*E. fuscus*), little brown bats (*M. lucifugus*), northern long eared bats (*M. septentionalis*), eastern small footed myotis (*M. lebeii*), Indiana bats (*M. sodalis*), and tri-colored bats (*P. subflavus*) are hibernating species, often found in large cave hibernacula during the winter months (Whitaker *et al.*, 1998). The big brown has been thought of as a habitat generalist, although some preferences have been documented at the local level (Agosta, 2002). Little brown bats forage over ponds or in forests and northern long eared bats prefer to forage over forested hillsides and ridges (Harvey *et al.*, 2011). The northern long eared bat is currently being listed as threatened by the USFWS.

Data on foraging habits of the eastern small footed myotis is sparse, but examining its dietary preferences (flies, mosquitoes, true bugs, and other insects) indicates they may be habitat generalist in foraging. Indiana bats have been recognized as an endangered species for decades, and research has pointed to riparian and floodplain foliage as foraging habitat (Humphrey *et al.*, 1977). The tri-colored bat has been noted as a species that "forages over waterways and in forests" (Harvey *et al.*, 2011).

The New York Department of Environmental Conservation (NYSDEC) started a statewide active, vehicular acoustic monitoring program in 2006 in order to better asses migratory, "tree dwelling", bat species populations. Historically, "cave dwelling" bat populations have been better studied, largely due to the greater ease and logistics of winter cave assessments of hibernating species. Around the same time the monitoring program was initiated, WNS was discovered. The program was quickly adapted to include all species of bats, in an effort to monitor the impacts of WNS. To date the NYSDEC has collected 5 years of data, with almost 10,000 km of surveys, all of which has been done largely through volunteer staff.

Using the NYSDEC's outlined protocol (described in Appendix 1), we conducted bioacoustical surveys in the summer of 2013 to assess bat species diversity in Chautauqua County and the relationship that habitat type and various habitat characteristics have on bat activity. There are three main areas of focus for this study; calling activity interactions with habitat, environmental variables, and time.

It is hypothesized that bat species will utilize habitat types differently based on a species' morphology and habitat type. Bats that are larger will forage in less spatially cluttered habitats versus smaller species, so calling activity will be higher for larger species of bat. Additionally, calling activity will be affected by habitat type. The hypothesis is that some habitat types are more beneficial in terms of nutrient resources than others. For example, forested and riparian areas should be a greater foraging resource than agricultural habitats based on insect populations present. Agricultural fields are generally treated with pesticides and insecticides, and as such should have lower numbers of prey, which translates to lower levels of bat calling activity.

    Environmental variables examined in this study include elevation, and soil type. It is hypothesized that there will be a relationship between bat call activity and elevation. Lower elevations are more consistent with bodies of water and higher temperatures, so will exhibit higher levels of bat call activity. In terms of the hypothesis regarding soil type, it is posited that soil types more consistent with forested (well drained gravelly or channery loam) and riparian (poorly drained muck or silt loam) will experience higher levels of bat call activity.

    Last, time of the summer will have a relationship with levels of bat calling activity. The emergence of young pups as flying adults, as well as fall swarm events associated with migratory species, will cause peaks of calling activity at different points in the summer, specifically in July and August.

Methods and Materials

Habitat

Prior to acoustic sampling a thorough habitat analysis was conducted along the 34 km of the Jamestown transect as well as the 18 km of the Arkwright transect. Habitat types were categorized as forest, scrub, wetland, meadow, agricultural, or residential; along with hydrologic types such as streams and lakes/ponds. Forest habitats were defined as any region that was completely covered with tree species and less than 10% of open canopy. Wetlands were designated as such based on the vegetation found at the site (United States Army Corps of Engineers Delineation Manual) or identified through GIS mapping resources. Scrub habitat was defined as an area 75% covered by shrub species (*Lonicera*, *Cornus*, *Malus* etc.), younger trees, and open canopy; however some mature tree species may still be present. Meadows were areas with herbaceous vegetative cover such as grasses, and less than 10% of woody plant species. Agricultural habitats were deemed as such if there was evidence of current agricultural practices being conducted, but excluded pastures which were listed as meadow. Residential habitat was an area of human usage, usually found as a house, parking lot, or barn, but was also expanded to include an adjacent yard or driveway. Relative frequency of habitat types was evaluated to ensure that there was no bias inherent in the transects. This was done in ArcGIS by reviewing the habitat type along the transects at every 0.4 km. Table 1 summarizes the habitat type definitions.

Table 1. Summary of Habitat Type Definitions. Included are type, definition explaining habitat type, and the relative frequency of occurrence of habitat types in the transects were adjusted to account for the co-occurrences of habitat types.

Habitat Type	Definition	Adjusted Relative Frequency
Forest	100% of habitat is mature tree species	22%
Wetland	Characterized by wetland flora species, standing water and/or hydric soils	13%
Scrub	75% of habitat is shrub species (crabapple, sumac) and immature tree species with some open areas of meadow	15%
Meadow	At least 90% of habitat is free of any tree or shrub species, and consists largely of grasses	22%
Agricultural	Evidence of recent use as an agricultural field	7%
Residential	Area of human habitation or construction	13%
Stream	Within 100m of road for larger streams or 10m for smaller	3%
Lake/Pond	Within 100m of lake or pond	3%

Habitat was determined through use of a Garmin GPS (Garmin International, Inc. Olathe, KS) handheld unit. Each transect was delineated on foot or in a vehicle, with separate waypoints recorded each time the habitat changed, or if the road passed over or near a body of water (lake, pond, stream). This helped in determining habitat associated with a specific bat call, however discrete measurements were still used in GIS to accurately designate habitat type. The habitat on each side of the road was noted by its direction therefore a North-South road habitat was noted as east or west, and an East-West road was noted as being north or south. The resulting waypoints were uploaded into ArcMap's Geographic Information System (ESRI, Redlands CA). An attribute table was created, with the waypoints being further outlined by entering the specific habitat data for each point recorded. As each waypoint could potentially represent several habitats, the attribute table in ArcMap reflects this (for example there may be a forest habitat on the

North side of the road, and a wetland habitat on the South side and this would have been entered into the attribute table as "Forest-Wetland").

During habitat analysis the bat calls identified to species were matched with the habitat category specific to their location. Other data based on paper maps, GIS data sets, and aerial orthoimagery (NYS GIS Clearinghouse - NYS ITS GIS Program Office, Albany, NY) was also used in this process. Bat species were considered in a wetland, lake, pond, or stream habitat if they were within 100 m of the boundary (Schirmacher *et al.*, 2007) based on GIS wetlands maps and/or actual visible boundary of lake/stream. For small, perennial streams a distance of 10 m was used. The difference in approach for smaller streams was based on the idea that such streams are a significantly smaller foraging resource and as such would have a more limited nutrient value in the landscape. The criterion above allowed for a repeatable standard in habitat use determination.

Bat Surveys

NYSDEC acoustic transect protocol was used for survey methodology. Bat surveys were conducted using Binary Acoustic Technology's Model AR125 (Binary Acoustic Technology, Tuscon, AZ) ultrasonic microphone, retrofitted with a protective case and a magnetic bottom in order to attach to the roof of an automobile. Binary Acoustic Technology's SPECTR III (Binary Acoustic Technology, Tuscon, AZ) software was used to record the bats calls, and DeLorme's GPS (DeLorme, Yarmouth, ME) software was used to navigate and continuously record geographic coordinates. This allowed longitude and latitude to be determined for each individual bat call. Prior to the start of a survey environmental data including temperature, wind, precipitation % cloud cover and direction of the survey's start was recorded. Start points for each survey were reversed each time using the opposite direction of the previous survey (i.e. if a survey was started on the north end of the transect the next survey would start at the southern end). We did this to offset any temporal variation in bat activity. Surveys were conducted thirty minutes after sunset, on evenings that were 13 °C or greater at the start of the survey, had steady wind speeds below 24 kmph and little precipitation. This was done to ensure maximum calling activity of bat species and followed NYSDEC protocol. The vehicle was driven at a speed of 29 - 32 kmph, to match a bat's flying speed. This helped to ensure that each bat recorded was a separate individual, which has been a confounding factor in stationary acoustic studies.

Two transects were used including the Jamestown transect which runs from Bear Lake to the outskirts of Falconer, NY following Rt. 380. This is acoustic survey route that the NYSDEC uses for their annual survey. The Arkwright transect follows Rt 83, Zaihm Rd, Straight Rd and Rt. 79. Both transects are located in Chautauqua County. The Jamestown transect was selected by the NYSDEC for the varied habitats and areas of water (Bear Lake, Cassadaga Creek etc.) The Arkwright transect was selected for its comparable habitat types such as the areas of wetlands surrounding the Great Mud Lake, forested regions, and differing attribute of higher elevations. Twenty acoustic surveys were conducted during the summer of 2013 from May 16 to August 29. Ten surveys were conducted for each of the two transects. Surveys were attempted twice a week, one for each transect, however weather and temperature constraints (specifics noted above) did have an impact on survey frequency.

Species Identification

NYSDEC protocol was used to identify bat calls to species. Calls recorded using the SPECTR III software program were run through the SCANNR software program (Binary Acoustic Technology, Tuscon, AZ). The SCANNR program separates "noise" and incomplete bat calls from the rest of the files. Noise could be a result of a vehicle passing during the bat surveys, from wind, or biological sources, such as katydid calls. An incomplete bat call was not usable for species identification purposes, but was included in the overall call count. A recorded sequence of at least 5 or more sonar calls is necessary for accurate species identification (except for hoary bats, see below). In instances where multiple bats are recorded at the same time during a survey the following guideline was used. If the bat calls are separated by one second or longer, it is determined to be produced by a separate individual.

As mentioned above search phase calls were used for species identification (reasons given in Introduction). A search phase call is about 100 ms in duration, and depending on species are either "constant frequency" (CF), or "frequency modulated" (FM). A CF call would have little to no difference in frequency through the duration of the call, while a FM call could vary in frequency by quite a large range.

A first step in species identification was determining if the call is CF or FM. In New York State only the hoary bat shows CF call characteristics, therefore even though they rarely emit more than 4 sonar calls in a sequence, they are easily identifiable by their unique spectral pattern. Besides CF/FM comparisons, there are other key characteristics used to make a species level identification. One is the characteristics of the calls start, end and spectral pattern. Different species may have unique starting frequencies of their call or may have unique frequency patterns over the duration of the call. Unique frequency patterns include a downward or upward "hook" or "knee" shape to their call and where in the call and at what frequency it occurs is characteristic of a species. Another frequency characteristic used for identification is the minimum frequency of the call. *Myotis* species are problematic in species determination by their sonar calls; however the slope of the call has utility. The slope of the call refers to how steep the call is, and is a key characteristic used for identification. These differing aspects of a bat's sonar calls were used in assigning a call to a specific species through Discriminate Function Analysis (DFA) in R programming language (R Foundation for Statistical Computing, Vienna, Austria) and a comprehensive call library from the Eastern United States (Britzke 2003, Britzke *et al.*, 2011).

    Based on the repeatable results of past NYSDEC surveys this method has been deemed appropriate for acoustic studies, however, it does have limitations. Silver haired and big brown bat calls are virtually indistinguishable based on the ability of the species discrimination software, and given the relative rarity of silver haired bats, it is assumed that the majority of the calls are from big brown bats. Little brown, northern long eared, and tricolored bats are also readily detected by these acoustic surveys. Given the limited distribution and rarity of Indiana and small footed myotis species most of the DEC's survey routes would not be expected to encounter these species. While small footed bat is identifiable through the call analysis, the Indiana bat is often confused with other *Myotis* species. Hoary and eastern red bats are identifiable in numbers sufficient to track any changes in abundance (NYSDEC SWG Final Progress Report & Evaluation, 10/1/12 -

09/30/13). Table 2 notes the typical values for bat call components used in species identification.

Table 2. Typical Parameters for Identification of Bat Sonar Calls in NYS. CF refers to "constant frequency" calls type, FM refers to "frequency modulated" call type. The minimum frequency is the lowest measurement in a bat's call. Slope is the literal slope of the sonogram, recorded in octaves per second.

Species	CF/FM	Min Freq (kHz)	Slope (oct/s)
Big Brown	FM	22.5 - 32	n/a
Silver Haired	FM	22.5 - 32	n/a
Eastern Red	FM	32 - 38	<60
Hoary	CF	<22.5	n/a
Small Footed	FM	>45	n/a
Little Brown	FM	38 - 42	60 - 110
Northern Long Eared	FM	> 40	> 200
Indiana	FM	> 40	110 - 200
Tricolored	FM	> 40	< 100

Geo-spatial Analysis

Geographic coordinates were recorded simultaneously with the acoustic surveys. The ultrasonic microphone records bat calls, and includes a timestamp in the .wav format file that the call is recorded in. Time is recorded in different formats (UTM for the GPS data and EST for the acoustic data) however, it is possible to manually match these two files and obtain latitude and longitude coordinates for each call recording. In instances of multiple bat calls for a recording the specific coordinates are used for each separate call identified. In order to make the process more streamlined the NYSDEC's Natural Heritage program wrote a script in the statistics program R that automates the process by converting the times to the same format, matching up the bat calls with their respective coordinates, and importing this data into an excel spreadsheet.

When the process in R is complete, a data file is exported that contains the bat calls matched with each coordinate. This can then be imported into ArcMap. At this point in the GIS there are: waypoints outlining the bat survey transect that correspond to a specific type (or types) of habitat, and individual points that represent separate bat calls. These points are further stratified to illustrate the various species identified. After the specific coordinates for each identified bat call were imported into ArcMap they were able to be linked to elevation through a Digital Elevation model (DEM), soil type through a joins and relate process, and proximity to a hydrologic source by buffering and measuring each bat call's distance to a wetland, stream or lake/pond, which assisted in determining which habitat type a bat call was recorded in.

After a DEM was produced it allowed for a visual representation of bat calls in association with elevation gradients, as well as exported a specific elevation point in meters for each bat recorded. Each bat recording has a corresponding elevation point and we averaged these points for statistical analysis. Significant relationships were evaluated using the specific values produced from this DEM. Also, the actual map assisted in evaluating whether bat activity occurs in clusters at distinctive elevation points. Soil data analysis was conducted in much the same way as elevation data analysis. Using the United States Department of Agriculture's (USDA) soil data in GIS, a specific soil type was exported for each bat call recorded. A code for soil type accompanied the data, for statistical analysis.

Statistical Analysis

In order to ascertain how differing habitats were related to bat calling activity a loglinear analysis in SPSS (IBM Corporation, Armonk, NY) was conducted. All habitats were used as categorical variables, as well as species and temporal attributes (date). A Chi-squared test was conducted to determine whether the interactions with habitats were greater or less than the expected values of activity based on chance alone. For the Chi-squared test only calls associated with a single habitat value were used to avoid lack of independence of data from co-occurrence of habitats. Stream and lake/pond habitats were calculated separately to avoid a similar issue in that there were no instances of purely stream or lake/pond habitat, every call recording had an association with at least one other habitat. The associations between bat calling activity, soil type, and elevation were evaluated with separate univariate ANOVA tests in SPSS. Soil type and elevation were used as factors and bat call activity as the dependent variable.

Results

Habitat delineation provided 8 general habitat types including forest, wetland, scrub, meadow, agricultural, residential, stream and lake/pond. Adjusted relative frequencies of habitat occurrence were reported in Table 2. The relative frequencies were adjusted to reflect the frequent overlap of habitat types along the transects. The measurement indicates how frequent a specific habitat type was on the landscape. This not only allowed for identification of a bias in the transect, but also determined the expected values for bat calling activity. Calling activity differed from what was expected by occurrence of the habitat types in the environment ($X^2 = 455.2$, df = 5, $p < 0.0005$). Expected values and observed values used in the Chi-squared test are shown in Figure 1. Calling activity differed between stream and lake/pond habitat types ($X^2 = 25.1$, df = 1, $p < 0.0005$).

Figure 1. Observed versus Expected Habitat Type Frequency as Compared With Bat Calling Activity. "E" refers to the expected value of call activity based on frequency of habitat types in the environment, while "O" refers to the actual observed value of calling activity during the study.

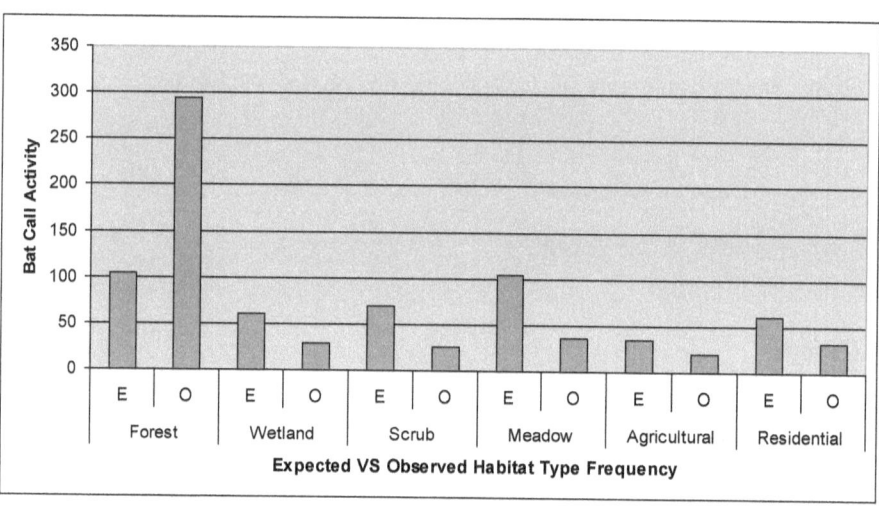

All 9 species of bat present in NYS were recorded, although there are discrepancies in accuracy of acoustic identification for some of the species, such as the Indiana bat, as well as issues of sample size for some of the rarer species recorded. Total number of calls that were identified is described in Table 3. For the purposes of this study the sample sizes of several species was insufficient to rely on for results. As such the following bat species were excluded from this discussion: small footed, northern long eared, Indiana, and tricolored. The overall number of bat calls in a specific habitat type will be discussed however.

Table 3. Bat calls identified from summer 2013 acoustic surveys and associated sample size.

Species	N
Big Brown	343
Silver Haired	399
Eastern Red	181
Hoary	241
Small Footed Myotis	2
Little Brown	52
Northern Long Eared	7
Indiana	5
Tri-Colored	17

A total of 1513 bat calls were recorded, of those, 1248 were able to be identified to the species level representing an identification success rate of 82.4%. The success rate of the Arkwright transect was lower than the Jamestown transect (78.4% vs. 84.5%), this is possibly due to acoustic interference from insect calling or dirt/gravel roadways that were more common in the Town of Arkwright.

Loglinear analysis was used to assess the interactions between species, habitat, and time, which revealed significant interactions between habitat and species (see Table 4, Figs. 2 & 3), time and species, as well as among habitats (see Table 5). The goodness-of-fit test resulted in a non-significant Likelihood Ratio ($p > 0.9$), indicating a well working statistical model. A best fit model is the outcome of the analysis that dropped non-significant interactions ($p > 0.05$) from the model to end up with the significant effects ($p < 0.05$). This process continues until all of the remaining interactions are significant. All interactions were a function of bat calling activity, as each measurement was associated with a specific recording of a bat from the acoustic surveys.

Table 4. Significant Interactions Between Species and Habitat Type. Four habitat types showed significant interactions in loglinear analysis. The overall percentage of calls per habitat type is shown, as is the % by the total N, and the percentage by the specific habitat type.

Forest (df = 9, p = 0.001)	Species	N	% By Total N	% By Habitat
896/1272 = 70.4%	EPFU	256/343	74.63	28.6
	LANO	296/399	74.18	33
	LABO	135/181	74.58	15.1
	LACI	150/241	62.24	16.7
	MYLU	22/52	42.3	2.5
Wetland (df = 9, p = 0.01)				
340/1272 = 26.72%	EPFU	114/343	33.23	33.5
	LANO	115/399	28.82	33.8
	LABO	30/181	16.57	8.8
	LACI	51/241	21.16	15
	MYLU	16/52	30.76	4.7
Residential (df = 9, p = 0.01)				
241/1272 = 18.94%	EPFU	55/343	16.03	22.8
	LANO	68/399	17.04	28.2
	LABO	28/181	15.4	11.6
	LACI	64/241	26.55	26.6
	MYLU	13/52	25	5.4
Stream (df = 9, p = 0.001)				
164/1272 = 12.89%	EPFU	33/343	9.62	20.1
	LANO	76/399	19.04	46.3
	LABO	20/181	11.04	12.2
	LACI	30/241	12.44	18.3
	MYLU	3/52	5.76	1.8

Figure 2. Bat Calling Activity as Compared Across Species and Habitats. All habitat types were compared with the 5 species that had sufficient sample size for discussion.

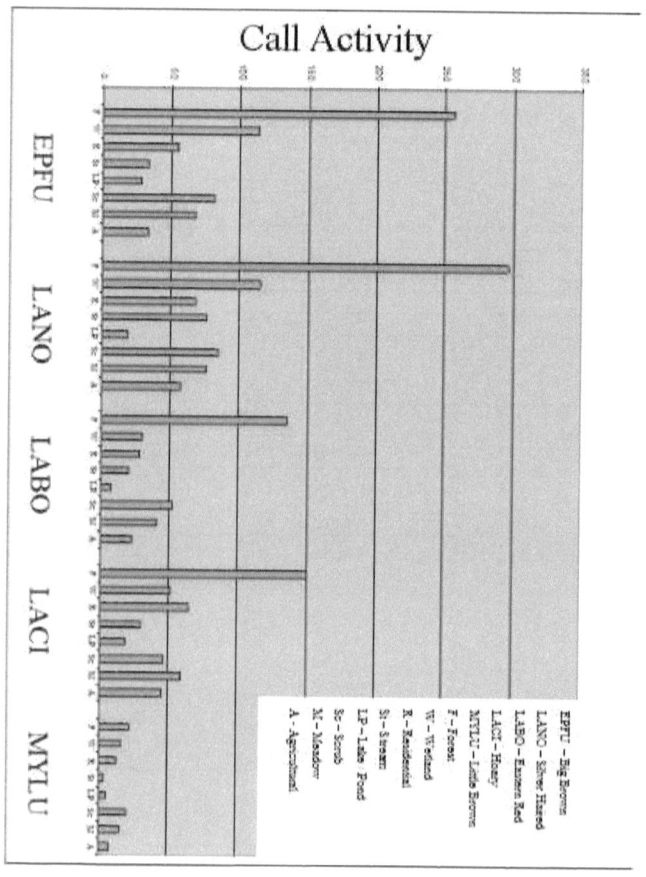

Figure 3. Bat Call Activity and Habitat Type. The habitats that had significant interactions through loglinear analysis are compared with the species that had sample N > 50.

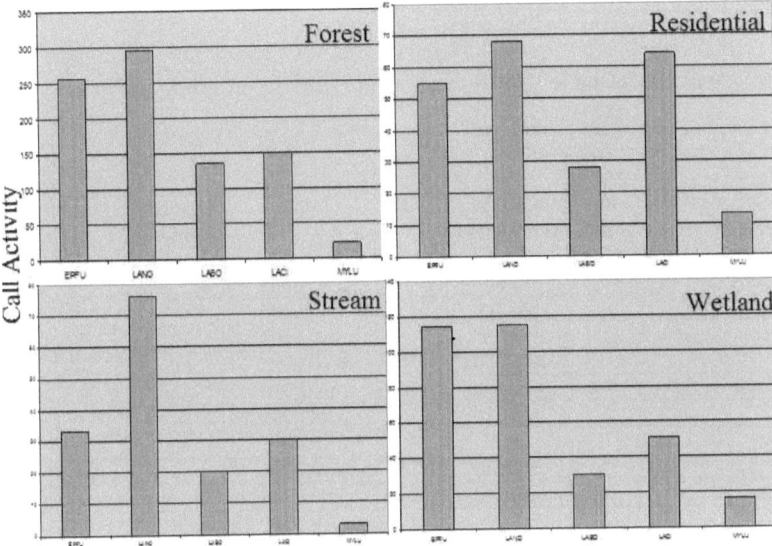

Big brown, silver haired, and eastern red bats were all very close (~75%) in frequency in forest habitats (calculated as a percentage of total N of each species). Big brown and silver haired bats were again close in frequency of call activity in wetland habitats, while hoary bats had the highest frequency in residential. Silver haired bats had the highest frequency in stream habitats in terms of both call activity per species and per habitat.

There were many (26) significant (p < 0.05) habitat to habitat interactions. Table 5 outlines the minimum and maximum for each significant habitat type, in terms of percentage of shared habitat.

Table 5. Significant Habitat to Habitat Interactions. The interaction results for maximum and minimum shared habitat, % of shared habitat, and overall trend between habitats are shown below.

Habitat Type	# Of Bat Calls	Bat Calls (%)	Habitat Interaction	df	Sig	Shared Habitat (%)	Trend
Agricultural	176	13.8	Forest	1	$p < 0.0005$	47.2	+
			Stream	1	$p < 0.0005$	0	-
Forest	896	70.4	Wetland	1	$p < 0.0005$	22	+
			Lake/Pond	1	$p < 0.0005$	4.2	-
Lake/Pond	85	6.7	Residential	1	$p < 0.05$	48.2	+
			Stream	1	$p < 0.0005$	0	-
Meadow	274	21.5	Scrub	1	$p < 0.0005$	26.3	+
			Stream	1	$p < 0.0005$	2.6	-
Resident	241	18.9	Wetland	1	$p < 0.0005$	22.8	+
			Stream	1	$p < 0.0005$	2.9	-
Scrub	292	23	Stream	1	$p < 0.0005$	1.7	-
Stream	164	12.9	Wetland	1	$p < 0.005$	17.1	-

Agricultural, forest, lake/pond, meadow, resident, scrub, and stream habitats had significant interactions with other habitats as a function of bat activity. Agricultural habitats (forest: df = 1, p < 0.0005; lake/pond: df = 1, p < 0.05; stream: df = 1, p < 0.0005; wetland: df = 1, p < 0.005; meadow: df = 1, p < 0.0005; scrub: df = 1, p < 0.0005; and residential: df = 1, p < 0.0005) had the largest number of habitat-habitat interactions. Forest also had significant habitat-habitat interactions (wetland, scrub, meadow, residential, lake/pond: df = 1, p < 0.0005); as did lake/pond (meadow: df = 1, p < 0.0005; wetland: df = 1, p < 0.05; scrub: df = 1, p < 0.0005; residential: df = 1, p <

0.05; and stream: df = 1, p < 0.0005); meadow (wetland, scrub, residential, and stream: df = 1, p < 0.0005); residential (wetland, scrub and stream: df = 1, p < 0.0005), scrub (stream: df = 1, p < 0.0005); and stream (wetland: df = 1, p < 0.0005). These interactions are based on coincidence greater than one would expect by chance, or are a result of greater calling activity when the habitat types co-occur.

When comparing temporal variation among transects, there were significant interactions between date and bat species (df = 135, p < 0.0005), as well as date and habitat type (agricultural: df = 15, p = 0.01; lake/pond: df = 15, p = 0.004; residential: df = 15, p < 0.0005; stream: df = 15, p < 0.0005, and wetland: df = 15, p = 0.004). In terms of interactions between species, the call data show noticeable peaks in number of bats recorded per transect during the end of July. This varied by species. On average, big brown and silver haired bats dominated the transects in terms of bat calls recorded per week, followed by eastern red and hoary bats. *Myotis* species, along with the tricolored bat, had generally sporadic encounters along the transect, and their sample sizes were often too low to infer any interactions.

There was a peak of activity during the first and second week of June, again in similar time of the month in July, and again towards the end of August. Similar patterns of significance followed for lake/pond, stream, and wetland habitats. Of the habitat-time interactions, only stream and wetland habitat types were additionally significant as a habitat-bat call interaction.

The univariate ANOVA of calling activity and elevation resulted in significance for elevation as a factor (F = 4.899, df = 9, p < 0.0005) as well as for calls associated with certain species. Call activity was significant for big brown (p = 0.034), silver haired (p =

0.006) and little brown (p = 0.042) bats. Figure 4 illustrates the trend in bat calling activity with elevation. The univariate ANOVA of calling activity and soil type did not yield statistically significant results (p > 0.05).

Figure 4. Histogram illustrating calling frequency with elevation.

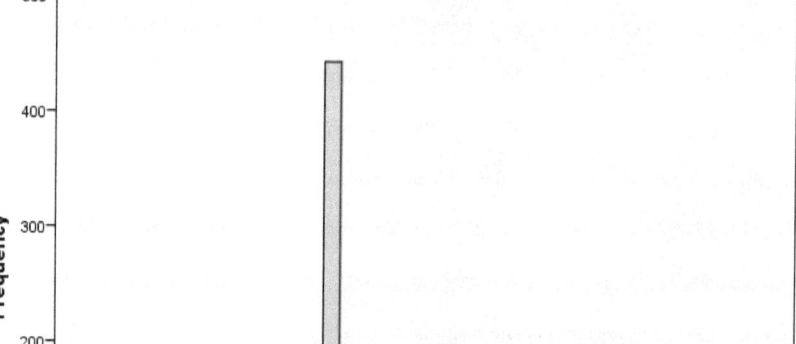

Discussion:

This experiment showed a strong relationship between bat activity and forested habitats, as well as species specific patterns of habitat usage. Additionally, we discovered that bat species are not foraging exclusively in one habitat type, that there are strong relationships but not exclusive ones. Elevation showed strong statistical significance, especially for big brown, silver haired, and little brown bats. As stated earlier, their collective average elevation measurements varied by just 4 meters indicating that elevation could potentially be used as a predictor variable for recording bat species as well.

Species and habitat interactions revealed some surprising results. While it may seem intuitive that forested areas are important for bat species (especially tree roosting species), the resulting data showed that more than 2/3 of the calls recorded had forest as a component of their habitat, and almost 1/4 of bat calls were recorded only in forests. This indicates that forests are not only crucial to roosting species, but given our recording timeframe, are a critical foraging resource as well. Forest habitats were significant for big brown, silver haired, and eastern red bats. Hoary, and little brown bats also had strong relationships with forests. What was surprising was the relatively low numbers of foraging bats around streams and lakes/ponds. This may be a bias in the transect, however, when looking at the relative frequency of habitat occurrence this was largely not an issue. It may be that to accurately assess foraging behavior of bats along streams and over lakes/ponds one needs to survey directly over these areas.

Wetlands were another habitat variable that showed a strong relationship with bat activity. More than a quarter of all the bats recorded during the surveys had wetlands as a component of their foraging habitat, and this was significant to big brown, silver haired, and little brown bats; although again eastern red and hoary bats also had relatively large numbers in these habitats.

Residential and stream habitats were the other two habitat types that showed significant results upon the completion of loglinear analysis. However, when looking at the species profiles it appears that these measurements may have been biased by the low sample size of the more uncommon species. For instance, with residential habitat most of the species had frequencies under 20%, with the exception of hoary and little brown bats. Northern long eared (N = 7) and Indiana (N = 5) bats had 42.85% and 80% of their sample size in a residential area, respectively. This resulted in significance from the loglinear analysis, but when factoring in the low sample sizes it is possible that these two species biased the overall measurement. Hoary bats had 26.5% and little brown bats had 25% of their sample sizes associated with residential areas, which is interesting, and may support the hypothesis that as a large species they tend to seek out a structurally uncluttered environment. In the case of the little brown bat it may indicate that they are foraging generalists, as review of their distribution across habitats also indicates.

The significance attributed to stream habitats may also have been skewed by low sample sizes. In this case, two species, the small footed and Indiana myotis had zero call recordings in stream habitats. Again, the analysis computed this as significant because it was a strong negative trend. The other species in stream areas all had frequencies less than 20%.

The Chi-squared test showed that forest habitats ($X^2 = 344.3$) had nearly triple the amount of calls one would expect based on chance, while the other habitat types all had calling activity lower than expected. This was especially true for agricultural habitat ($X^2 = 6.3$), supporting the hypothesis that agricultural habitats have a negative relationship with bat calling activity. However, the test was run with only single habitat interactions to ensure independence in habitat types, loglinear analysis indicated that habitats could be enhanced as a foraging resource based on proximity to a more nutrient rich habitat type (Table 5). The Chi-squared test for stream and lake/pond habitats indicated that despite equal representation on the landscape (adj. rel. freq. = 3%, $X^2 = 12.53$), the observed values for streams were well above what was expected by chance. The opposite was true for lakes/ponds (Figure 1).

The temporal component of the experiment revealed that there are indeed peaks in bat foraging behavior during specific parts of the summer months. As hypothesized, there were larger numbers of bats recorded during the time that young of the year start to fly, as well as at the end of August when fall migration begins. Temporal - habitat interactions revealed interesting results. While the habitat types themselves were not directly significant to species specific measurements, there was most assuredly a relationship between temporal variables and habitat types. For example, more bats were recorded in agricultural areas in the 5th, 9th, 12th and 14th weeks of the study, which corresponds to early June, early July, late July and the middle of August. Temporal variation was also not significant for forest habitats, suggesting that there is not necessarily a most active time of the summer for foraging. It may be that forest habitats are simply excellent resources for foraging bats no matter the time of year.

Temperature could have impacted the number of bats recorded per night; however this was controlled for in the protocol whereas survey nights that had temperatures below 13 °C were cancelled. Another potentially confounding factor when interpreting the results is that some weeks had more surveys than other weeks. This issue is alleviated by focusing on the relative frequencies of bats recorded as opposed to the actual numbers.

The numerous significant habitat-habitat interactions can be explained by considering two things. First, the data analyzed was initially a function of a single recorded bat call. Therefore, significant habitat to habitat interactions may indicate that many bat species do not forage in one habitat continuously. Bat foraging habits may be dynamic over the course of the summer, week, even day to day. Second, the habitat values recorded were often more of a checkered matrix of types than a distinct gradient. Some habitat types were strongly associated with others, for example forest and wetland habitats shared boundaries quite often while agricultural and stream habitats did not have a single co-occurrence. There were several instances of significant habitat to habitat interactions, indicating that this patchwork of landscapes contributes to more opportunities for bat species to find their particular prey of choice, and reinforces other studies findings that bat species, as flying mammals, are not as detrimentally affected by habitat fragmentation as others (Law *et al.*, 1999).

Overall this study reinforced the strong link between bats and forests. While it did not reveal significant results for bodies of water such as lakes and ponds, this may be a limitation of the vehicular survey methodology, and not necessarily indicating a lack of significance. Perhaps a roadside survey is not adequate to describe bat activity along waterways. This can be addressed by boating based acoustical surveys, which would also be an interesting experiment. It may also be that a flying bat can easily reach a water source, and as such has no need to forage in the immediate area of one. Additionally, the results suggest that bats are not attracted to a monoculture of habitats, that foraging behavior is widely varied and can involve many different habitat types and interactions. Of particular interest is the spike in silver haired bat activity in stream habitat. As stated earlier, big brown and silver haired bats face issues with acoustic identification, so overall the numbers of each species recorded is in question. However, there is a noticeable spike in silver haired bat activity around stream habitats, indicating that these numbers may at least provide some information on foraging behavior, even with species that are problematic in acoustic identification.

Finally, the study helped to illustrate the utility of mobile acoustic surveys. The methods used were sufficient to identify at a coarse, broad level, areas of high bat activity. This is crucial in reducing the amount of time spent sampling large areas extensively, and could reduce costs as well as enhancing the targeted area or species. While the results seem to merely support preexisting data on foraging ecology, the reader should bear in mind that information on multiple species of bat from over 48 km of survey area was surveyed for in one study. Previous studies were constrained to focus on single species or groups of species, or more alternately a specific ecological gradient. The utility of the protocol used can be applied to various ecological questions, while adding to the existing pool of data.

Acknowledgements

I would like to thank my thesis committee members Dr. Karry Kazial, Dr. William Brown, and Dr. Jonathan Titus for their invaluable assistance, advice, and the countless hours they spent helping me perfect this project. Also, I would like to thank Dr. Ann Deakin and Professor Andrea Austin for their help with geosciences and statistics, respectively. Equipment, training, and species identification expertise was provided by the NYSDEC, this thesis project would not have been possible without the assistance from that organization, specifically Carl Herzog and Kathleen O'Connor. Volunteer assistance from Jessica Pasieka was instrumental and her time and help was greatly appreciated, this research would not have been possible without her continual availability and support. Finally, I would like to sincerely thank the Holmberg Foundation for funding my research.

Literature Cited

Agosta S.J. 2002. Habitat use, diet, and roost selection by the Big Brown Bat (*Eptesicus fuscus*) in North America: a case for conserving an abundant species. *Mammal Review.* **32**, 179-198

Baerwald E.F., D'Amours G.H., Klug B.J., Barclay R.M.R. Barotrauma is a significant cause of bat fatalities at wind turbines. *Current Biology.* Vol. 18 no. 16 R696

Brandes, T.S. 2008. Automated sound recording and analysis techniques for bird surveys and conservation. *Bird Conservation International.* 18:S163-S173

Britzke E.R. 2003. Uses of ultrasonic detectors for acoustic identification and study of bat ecology in the Easter United States: An abstract of a dissertation. *Tennessee Technological University*

Britzke E.R., Duchamp J.E., Murray K.L., Swihart R.K., Robbins L.W. 2011. Acoustic identification of bats in the Eastern United States: A comparison of parametric and nonparametric methods. *Journal of Wildlife Management.* 75(3): 660-667

Carlberg, V.E. 2009. Agriculture in Chautauqua County: A discussion paper. Cornell Cooperative Extension of Chautauqua County

Carter, T.C., Menzel, M.A., Owen, S.F., Edwards, J.W., Menzel, J.M., Ford, W.M. 2003. Food habits of seven species of bats in the Allegheny Plateau and ridge and valley of West Virginia. *Northeastern Naturalist.* 10(1): 83-88

Harvey, M.J., Altenbach, J.S., Best, T.L. 2011. Bats of the United States and Canada.

Humphrey, S.R., Richter, A.R., Cope, J.B. 1977. Summer habitat and ecology of the endangered Indiana bat, *Myotis sodalis*. *Journal of Mammalogy.* Vol. 58, No. 3, pp 334-346

Jones G., Jacobs D.S., Kunz T.H., Willig M.R., Racey P.A. 2009. Carpe noctem: the importance of bats as bioindicators. *Endangered Species Research.* Preprint 1-23

Klug B.J., Turmelle A.S., Ellison J.A., Baerwalk E.F., Barclay R.M.R. 2011. Rabies prevalence in migratory tree-bats in Alberta and the influence of roosting ecology and sampling method on reported prevalence of rabies in bats. *Journal of Wildlife Diseases* 47(1), 64-67

Kunz T.H., Braun de Torrez E., Bauer D., Lobova T., Fleming T.H. 2011. Ecosystem services provided by bats. *Annals of the New York State Academy of Sciences Issue: The Year in Ecology and Conservation Biology.* 1223, 1-38

Lacki, M.J., Hayes, J.P., Kurta, A. 2007. *Bats In Forests*

Law, B.S., Anderson, J., Chidel, M. 1999. Bat communities in a fragmented forest landscape on the south-west slopes of New South Wales, Australia. *Biological Conservation*. 88 (1999) 333-345

Loeb, G. 2012. Grape insect and mite pests – 2012 field season. Department of Entomology, Cornell University

Mankin, R.W., Hagstrum D.W., Smith M.T., Roda, A.L., Kairo, M.T.K. 2011. Perspective and promise: A century of insect acoustic detection and monitoring. *American Entomologist*. Vol. 57. No. 1. 30-44(15)

Medellin R.A., Equihua M., Amin M.A. 2000. Bat diversity and abundance as indicators of disturbance in neotropical rainforests. *Conservation Biology*. Vol. 14. No. 6, 1666-1675

New York State GIS Clearinghouse

NYSDEC SWG Final Progress Report & Evaluation, 10/1/12 - 09/30/13

Narvins, P.M. 1995. Frog Communication. Scientific American

Schirmacher M.R., Castleberry S.B., Ford W.M., Miller K.V. 2007. Habitat associations of bats in South-central West Virginia. *Proc. Annu. Conf. Southeast Assoc. Fish and Wildlife Agencies*. 61: 46-52

Soil Survey Staff, Natural Resources Conservation Service, United States Department of Agriculture. Web Soil Survey. Available online at http://websoilsurvey.nrcs.usda.gov/. Accessed 5/6/2014

United States Army Corps of Engineers Wetlands Delineation Manual

Whitaker, J.O.Jr., Hamilton, W.J.JR. 1998. *Mammals of the Eastern United States*

Yovel Y., Melcon M.L., Franz M.O., Denzinger A., Schnitzler H. 2009. The voice of bats: how greater mouse-eared bats recognize individuals based on their echolocation calls. *PloS Computational Biology* 5(6): e1000400. doi:10.371/journal.pcbi.1000400

Appendix 1 -

NYSDEC Acoustic Bat Survey Protocol:

INTRODUCTION 26 April 2011
Welcome to the New York's Bat Acoustics Survey!
FOR THE FASTEST ROUTE TO GET STARTED, SCAN THROUGH THE HIGHLIGHTED PASSAGES. This will give you a quick overview. After you get the big picture you can go back and look at the details, as necessary. Most of New York's 9 bat species are currently faced with threats of an unprecedented nature and scale. Your participation in this survey will help us keep track of the status of bat populations all across the State, providing important information that we simply cannot get any other way. Make no mistake: your help and that of others like you are essential to the success of this project. Many of you have heard of White-nose Syndrome, the mysterious condition that is killing vast numbers of bats as they hibernate in caves and mines during the winter. (If you would like to learn more, see http://www.fws.gov/northeast/white_nose.html). While we can keep track of some of the cave bat species by counting them while they hibernate in winter, others tend to hide out of sight in cracks and crevices. Also, although the very common big brown bat is considered a cave bat, we have never been able to account for more than a tiny percentage of the state's population with hibernation site surveys. It is thought that most may spend the winter in attics and other human structures, making access difficult. This survey will help us keep track of these species that are not well covered by winter cave surveys. Less well known than White-nose Syndrome is the threat faced by the three species of bats that fly south for the winter. This group includes the least frequently encountered bats in our region and yet they are by far the most common bats to be killed by wind turbines. The seeming rarity of these bats suggests that the numbers being hit by rotating turbine blades might form a significant percentage of the population, but we really don't have a very good idea. This study will help us to determine if migratory bat populations are doing well. Before we get into the nitty-gritty of how you will do the surveys, though, let me say something clearly. Although the results of this project are important to our ability to keep track of and hopefully ensure the continued existence of NY's bats, your personal safety comes first. Please obey all traffic laws and heed all of the safety advice given here. Although we are not asking you to do anything risky – it's probably less dangerous than if you drove a similar distance in your normal manner - we don't want to have even a single traffic safety incident. This activity is a bit unusual, though, in that the speed you will be driving at is 18-20 mph. At that speed, your chief safety concern is being overtaken by vehicles that may be moving faster than you. Because the routes have been chosen to avoid high-speed roads and traffic at this time of day is generally low, experience suggests that this is not a difficult situation to deal with. If you see that you are about to be overtaken, simply turn on your emergency flasher lights, pull over as far to the right as possible at some convenient spot, and stop until the vehicle passes. Turn off your flashers and resume the driving as soon as it is practical to do so. The trip report (which we'll discuss more thoroughly later) has a place for you to enter approximately how many times it was necessary to do this. For most of you it will be an infrequent occurrence. If it happens more often than you are comfortable with on any one night, it might be best to terminate the survey and notify your coordinator as soon as possible. Use your best judgment in this.

INSTRUCTIONS
Most of what follows is included in the accompanying video presentations. Just pop the DVD into most any Windows PC, find and open the video folder, and double click the appropriate video file.

Overview
Full details for every step are available later in this document. The detailed procedures are written for users of Windows XP. There may be minor changes if you have Windows 2000, Vista, or Windows 7.

At its simplest, this project involves attaching a bat detector to the roof of your vehicle, connecting it to a laptop PC and driving a predetermined route while the equipment records the calls of the bats you encounter. You will then send the recorded files to our office by mail on a USB thumb drive that we provide. The routes each take about an hour to drive.
Here's a step-by-step outline of what you need to do:
STEP 1: Verify that your computer meets the requirements
STEP 2: Install the SPECT'R software, used with the bat detector.
STEP 3: Install the Delorme Street Atlas software, which will tell you where to do the survey and keep track of where each bat that you encounter was found.
STEP 4: Connect the bat detector and ensure it is working
STEP 5: Load your route(s) onto the map
STEP 6: Connect the GPS antenna and ensure it is working
STEP 7: Perform the survey

STEP 1: Verify that your computer meets the requirements
This project will sample for bats from all areas of New York State. You can probably appreciate that managing a project of this scale can be a challenge in many ways, and unfortunately the budget for the study does not allow us to purchase a laptop computer for every participant. While we can provide all of the other equipment needed, our hope is that most participants will have access to a computer that can be used. Nearly all laptops manufactured in the last 5 years are suitable. Some that are even older than that will work fine as well.

System requirements include:

a. Microsoft Windows operating system: Windows 2000, XP, Vista, or Windows 7.
b. Pentium processor or equal, 1 GHz or greater clock speed
c. 128 Meg Ram (512 for Vista)

38

d. 800 MB free hard drive space for software installation; 3 GB is desirable (see discussion below)
e. DVD drive
f. Qty (2) USB 2.0 ports. NOTE: USB 1.1 ports found on older computers are too slow for the bat detector and will not work.
g. Correctly set date and time

How to check if you have USB 2.0
Recently manufactured computers will almost certainly have USB 2.0. If in doubt, you can check yours pretty easily:
• Right-click **My Computer**
• Click **Properties**
• Click the **Hardware** tab
• Click on **Device Manager**
• Scroll down until you see **Universal Serial Bus Controllers**, and click the "+" sign
• The exact wording will vary from one computer to the next, but if you see the word "enhanced", even once, when you expand the **Universal Serial Bus Controllers** line, then your computer has USB 2.0. Generally speaking, if one of your ports is USB 2.0 then all of them are. See figure 1, and the video file 1_USBCheck.wmv, available in the videos folder of the DVD. You will need 2 USB ports, one for the bat detector and one for the GPS.

How to check available hard drive space
Hard drive space could possibly be a problem for some people. In addition to the space necessary to install the software, it is easiest and safest to record the bat calls onto the hard drive. If you encounter a lot of bats then the space requirement to do this can be significant, perhaps 1 GB or more in extremely busy areas. If you have lots of spare space then there's nothing to worry about, though. Again, it's pretty easy to check : Just double click **My Computer**. In the **Hard Disk Drives** section you will see **Local Disk (C:)**. Expand the window or scroll to the right until you see the **Free Space** column associated with the C: drive. (For a video demonstration see 2_Hard_Drive_Check_and_Clock_Set.wmv, available in the videos folder of the DVD). If it shows 3.0 GB or more free space then you are good to go. If not, and you are at a loss as to how to free up some space, then refer to the appendix for some ideas. If this still doesn't work for you then contact us and we can work with you to figure out some way to temporarily free up enough space to make it work.

How to set your computer's clock
The software uses your computer's clock to keep track of things. It is important that the indicated date is correct and it's also helpful if the indicated time is accurate. It's pretty easy to set the date and time. Double click on the clock display (normally in the lower right corner of the screen). If the indicated date is incorrect, just click on the current date in the calendar. To change the time, highlight the appropriate digits in the time display and enter the correct numbers. Click **APPLY** and then **OK** to save the settings. (For video see 2_Hard_Drive_Check_and_Clock_Set.wmv in the videos folder of the DVD.)

SOFTWARE INSTALLATION
Software installation is a two-step process, one for the Bat Detector and the second for the GPS mapping. Both are supplied on a single disk. Perhaps you are concerned about the propriety of installing these licensed software products on your computer. If so, good for you! NY DEC takes software licensing issues seriously and you can rest assured that, operating as our agent on this project, you are legally entitled to install the supplied software. We do ask, though, that you uninstall the software at the end of your involvement with the project.

STEP 2: Install the SPECT'R software
The software used with the bat detector is called SPECT'R. A new installation is usually quite easy (See 4_ InstallSPECTR.wmv, available in the videos folder of the DVD). NOTE: If you have a very new laptop running one of the **64-bit variants of Windows** then you need to install a different version of SPECT'R. Here's how to tell if this applies to you:
• If your computer uses Windows XP or Windows 2000 then you do not have a 64 bit system. Skip to the installation procedure below.
• If you are running Windows Vista or Windows 7, Click **Start**
• Right click **Computer**
• Click **Properties**. Under "System", view the system type. If you see the phrase "64 bit" then you need to use the 64-bit version of SPECT'R, as described below. If you do not see the "64 bit" use the 32-bit version of SPECTR.

Uninstalling old versions of SPECT'R
NOTE: If you helped out on this project in last year and your computer still has the SPECT'R software installed then you are good to go – skip to step 3. We have made no changes to the SPECT'R software for this year. If you are using a computer that was used with a bat detector prior to 2009 and has an existing installation of SPECT'R, please go to the **Control Panel** and use **ADD/Remove Programs** to uninstall all previous installations of SPECT'R. The version of SPECT'R we are using is custom configured for this project. See 3_RemoveSPECTR.wmv, available in the video folder on the DVD.
Once all old versions of SPECT'R have been removed:
• Insert the NY Bat Acoustics Survey disk into your DVD drive.
• Double Click **My Computer**
• Double click the **DVD** drive.
• Double click the **SPECT'R** folder.
• If you have determined that your computer has a 64-bit operating system, double click the **64 Bit** folder and proceed as described below. If you do not have a 64 bit system, then simply proceed as described below.
• Double Click **Setup** (The one with file type "Application")
• Click **Next** to move past the Welcome window
• Click "I Agree" to accept the license agreement (on behalf of NYDEC) and click **Next**.
• Click **Next** to accept the default destination folder location

39

- Click **Next** to finish the installation
- Click **Close** to close the installer

STEP 3: Install the Delorme Street Atlas software
The GPS package we are using is DeLorme's Street Atlas. (For a video demonstration, see 5_DelormeInstall.wmv, available in the videos folder of the DVD.) If you helped out on this project in a prior year you should have uninstalled the Street Atlas software at the end of the project. If you didn't, don't worry about it - the version we are using this year is the same as in previous years.
- Insert the disk in your DVD drive (if it is not still there from the SPECT'R installation).
- Double Click **My Computer**
- Double click the **DVD** drive.
- Double click the **DeLorme** folder.
- Double Click **Setup.exe**
- At this time you may be asked to approve the installation of one or more prerequisite software items (Microsoft Visual C++ Redistributables, Microsoft DotNet, etc.) Click **Install** and wait until the process completes, usually a few minutes. If you don't see this, don't worry – it means you already have the required prerequisites to complete the installation.
- When the License Key Validation window appears, enter the 16-character license key on the DVD Case, 4 characters per box. It is not case sensitive. Click **Validate**.
- Click **Next** to launch the Install Wizard.
- Accept the license agreement (on behalf of NYDEC) and click **Next**.
- Enter NY in the **First Name** box, enter DEC in the **Last Name** box, uncheck the **Register Online** box, and leave all other fields blank. Click **Next**.
- Click **Next** to accept the default destination folder location
- Click **Next** to accept the default DeLorme Documents folder location
- The Custom Setup Window will appear. We will make 2 changes to the default setup.
    o Click the red **X** next to Activate Advanced File Management and then click **Install this feature**.
    o Click the box next to Automatically Check for Program Updates and then click **Do not install this feature**.
    o Click **Next**
- Click **Next** to turn down optimization for use with an Ultra-Mobile Personal Computer.
- Click **Install** to begin the installation.
- Wait patiently until the installation completes, usually several minutes, and then click **Finish**.

STEP 4: Connect the bat detector and ensure it is working
NOTE: You must do this well ahead of the time you expect to begin the survey. Connect the USB cable from the AR-125 Bat detector to an open USB port on the laptop. It doesn't matter which port you use, but it's easiest to remember which one it is and in the future always plug the detector into the same port. Your computer will supply all the power necessary to run the detector. Your computer should detect the new device and start a **Found New Hardware Wizard** to install the device driver. The directions that follow assume that you will use the Wizard. (For a video Demonstration, see 6_DriverInstallAndBatDetectorCheckout.wmv, available in the videos folder of the DVD.)
- If the Wizard asks you whether Windows can connect to Windows Update, select **NO** and click **NEXT**.
- The driver is located on the DVD. With the disk in the DVD drive, tell the Wizard to install the software automatically and click **NEXT**.
- The Wizard may warn you that the driver is not digitally signed. Ignore this and click **NEXT**. It may also warn you that the software has not passed Windows Logo testing. Again, ignore this attempt by Microsoft to control the entire world and hit **CONTINUE ANYWAY**.
- You may be notified that one or more files (ezusb.sys, ezxmt.sys, busbw2k.inf) are needed. If so, click **BROWSE** and navigate to the **DRIVER** folder on the DVD and then click **OK**.
- Click **FINISH** to close the Wizard.

If your computer does not automatically start the Wizard when you plug in the bat detector, the device drivers may already be installed. Proceed to the next step. Start the SPECT'R software package (double clicking the desktop icon is the easiest way). The SPECT'R window will open and should automatically detect the AR-125 bat detector. This will be indicated if the DEVICE line in the upper left of the window says AR125RevC or something similar. If it says Device: <No device found> then either the detector is not plugged in properly or is plugged into a different port than the one you used when you installed the driver. To test of the functioning of the detector, rub your thumb and forefinger together while holding them near the black face of the bat detector. This generates ultrasonic sound that the detector can hear and will be reflected in the display window. When you see the display change as you rub your fingers, you know it is working.
NOTE: There are a number of options available from the SETUP and WINDOW menus, but there is no need to change the default settings. If for some reason you suspect these settings have been accidentally changed, refer to the appendix for a list of the default settings.

STEP 5: Load your route(s) onto the map
If you helped out on this project in a previous year, please be aware that some of the routes may have changed for this year. This was mainly to fix problems such as closed roads, non-existent pavement, etc. Double click the Street Atlas desktop icon to start the GPS/mapping software. (For a video demonstration, see 7_DelormeInitialRouteLoad.wmv, available in the videos folder of the DVD.)
1) Check "Hide dialog and use current Map File Option at startup" the first time starting Street Atlas. The check box is in the lower left of the WELCOME window. Click **OK**.
2) Suggested: click HELP MENU (The yellow **Question Mark button**, in the tool bar at the top of the screen) and select "Shut off all pop-up tutorials".
3) Load your route file
    a. Click **ROUTE** tab, (bottom left)

b. Click **FILE** button, and then **IMPORT**
c. With the NY Bat Acoustics Survey disk in the DVD drive, navigate to the **ROUTE** folder on the DVD, click on the **[RouteName].anr** file, where [RouteName] is the name of the route you have been assigned to. Click **Open**.
d. The map display will show the entire route but the zoom level will likely be too far out to allow you to see all of the individual roads.
e. If you have been assigned multiple routes, repeat step c. for each route name.
f. The standard route begins at the end with the green dot and ends at the red dot. If you would prefer to start at the other end then repeat step c. above but choose **[RouteName] reverse.anr.** This will load the identical route but with the start and endpoints reversed.
g. Click the **SAVE** icon (**floppy disk**, upper left). Click **SAVE** to accept the default file name and location.
h. Zoom to approx 11-0 (**zoom-in button**, upper right – see figure 4). NOTE: If you zoom out too far then minor roads (which this study favors) do not display on the map. This is done to prevent crowding the map image with too many fine details.

4) Uncheck the **AUTO CALCULATE** and **AUTO BACK ON TRACK** check boxes, on the lower right (Figure 5).

STEP 6: Connect the GPS antenna and ensure it is working (For a video demonstration, see 9_GPSIntro1.wmv, available in the video folder of the DVD.)
1) Take the laptop outdoors and connect the GPS antenna to a different USB port than the one you used for the bat detector. When connected, a small red light on the antenna should begin flashing once every few seconds. Eventually this light will turn green, but there's no need to pay attention to this.
2) You may prefer to do the following from inside your vehicle. If so, place the GPS antenna on the roof as described below in the section titled "Mounting the Detector and GPS Antenna". If not using your vehicle, make sure that the yellow side of the antenna faces upward and that it has a largely unobstructed view of the sky.
3) Click the **GPS** tab (lower left, next to the ROUTE tab.)
4) Click **START GPS** (Lower left). The red GPS indicator on the lower right of the computer screen will begin to flash, indicating that the system is attempting to acquire satellites. Wait patiently for the system to acquire. It may take 10 minutes the first time you do it, but the GPS indicator will then turn green and stop flashing. A green dot will then appear on the map at your current location and the map will shift so this point is in the center of the screen. If you move from your current location, the dot will change to an arrow and the path you take will be drawn on the map. If after 15 minutes the GPS indicator on the lower right of the computer screen is still flashing red, consider the possibility that the view of the sky from your current location is not sufficiently open to allow for GPS function. Move away from obstructions to a clear area. If at any point the GPS indicator and/or the dot indicating your current position is yellow rather than green, this means that your position is being recorded, albeit at reduced accuracy. The accuracy in this condition is still adequate for this study, although in almost every case this is the result of an obstructed view of the sky from your current position, such as nearby large buildings, dense tree canopy, etc.
5) Once you have verified that the GPS is functioning normally, click **STOP GPS** (Lower left) to terminate the test. This action will bring up the "Save GPS Log File" window. Name the file "Test-MM-DD-YY" (where mm-dd-yy is the current date) and click **SAVE**.
6) Exit the Street Atlas program by clicking the red "close" button in the extreme upper right corner. You will be asked if you want to save changes. Click **Yes** to save all changes.
7) As a final check, you may wish to verify that there is still sufficient hard drive space for recording the bat calls.
   a. Double click **My Computer.** In the **Hard Disk Drives** section you will see **Local Disk (C:)**. Expand the window or scroll to the right until you see the **Free Space** column associated with the C: drive. If it shows 2.0 GB or more free space then you are good to go. If not, then refer to the Appendix. You have now verified that the software and hardware are functioning properly for the purposes of this study.

Uninstalling software after you are done
Uninstalling the Delorme and SPECT'R software when you are completely finished working on this project will free up disk space that you can use for other things in the future. It is also a requirement of our license agreement with DeLorme. (If you like the DeLorme program and want to have it for your personal use, consider purchasing a copy. It's quite reasonably priced; try Amazon.com.) Uninstalling is most easily done using the Add/Delete Programs function of the Control panel. In Windows XP:
1) Click START, SETTINGS, CONTROL PANEL.
2) Double Click ADD OR REMOVE PROGRAMS
3) Highlight the appropriate program (either DeLorme Street Atlas or SPECTR III) and click REMOVE. For a video demonstration, see 3_RemoveSPECTR.wmv, available in the video folder on the disk.

PERFORMING THE SURVEY
When to do the survey
Surveys must take place between May 28 and July 4. We must limit the allowable weather conditions in order to maintain the integrity of the survey process. Your local evening news broadcast is helpful here, and we recommend looking at the forecast for a few days prior to your planned survey night. It's best to take one last look at the forecast the night of the survey as well. The temperature at the beginning of the survey must be 55 ° F or greater. Although bats don't seem to mind a little precipitation, we need to pick a day when there is no chance of rain. The equipment can probably survive getting damp, but a severe drenching would do some harm. Also, rain drops hitting the bat detector will be much louder than the bat calls, making it impossible to get a good recording. Bats tend to avoid high winds and so we must avoid windy days for our surveys. If the planned evening has *steady* winds over 15 mph, wait for a calmer day. Signs that the wind is too strong are if leaves and smaller twigs on trees are in constant motion, paper is lifted from the ground and driven along, or you start to see whitecaps on larger water bodies. A gentle breeze is no problem, nor are occasional gusts. It's a strong, steady wind that we need to avoid.
All surveys begin 30 minutes after local sunset, approximately 8:30 to 9:30 PM, depending on the exact day and your location in the state. Please don't approximate the start time or look for the sun dropping below the horizon. These methods are not accurate enough for this study. Refer to the Start Time Calendar supplied in the equipment kit for a calendar of start times for each route and date.

If for some reason you don't have the provided table handy, you can easily calculate the start time for any given day by adding 30 minutes to a published sunset time, e.g., from your local newspaper, local TV weather reports, or by looking at web sites like: http://www.sunrisesunset.com/usa/New_York.asp. Remember to add 30 minutes to the sunset time in order to determine your start time. If you leave too early the bats will not be flying. Too late and you will miss the peak of activity. Please try to be punctual. NOTE: It will not be fully dark 30 minutes after local sunset. That's fine. It will be dark enough for the bats.

Setting up the vehicle
We have found that successfully running a new route is a two-person job. The passenger watches the laptop, keeps the driver on course and makes sure everything is working properly. The driver can then concentrate on driving and nothing else. Remember, safety is paramount. Caution: Sliding the bat detector assembly across the roof of the vehicle after the magnet is in contact may scratch the paint. When you wish to move the detector assembly tilt the entire assembly to break the magnet's attraction to the roof and then lift upward. For a video demonstration, see 8_DetectorMountAndHookup.wmv, available in the video folder on the DVD.

Mounting the detector and GPS antenna
Most people will find it convenient to have the cables from the detector and GPS antenna enter the vehicle through the passenger window. Simply open the window a half-inch or so. Please do not crush the cables by attempting to close the window all the way or by pinching them when shutting the door. An easy way to prevent this is to have the passenger in his or her seat with the door closed while the driver places the two devices on the roof. The driver then passes the cables through the slightly open window to the passenger. Both the bat detector and the GPS antenna are held on with permanently attached magnets. (Sorry, most convertibles won't work for this project.) The exact placement of the bat detector on the vehicle's roof is not critical, but it usually works well to position it approximately above the head of the passenger. Positioning the GPS antenna is similarly flexible, but it's best to separate it a bit from the detector. At least a foot or so in any direction is fine. Just ensure that it is placed on a mostly horizontal surface so that it faces generally straight upward. NOTE: People who see you driving with the bat detector in place will often assume that you left it there by mistake and warn you. Although this can be the start of an interesting conversation, please maintain awareness of your scheduled departure time. Both the GPS and bat detector cables have USB connectors for attaching to the laptop. Once you decide that everything is working during the pre-survey test (outlined above), plan to use that same configuration when you run the survey. If you change the configuration, the computer may think the bat detector is a new device and require you to install the drivers again. You could do this in the field, but will need to have the disk handy if it happens.

Powering the Computer
In 2009 we had all surveyors make sure that their computer was powered by plugging it into the vehicle's cigarette lighter using an adapter that we supply (the Xantrex 175 Plus inverter). This worked well for almost everyone, but with certain computers the results were problematic. For this year we are changing the protocol. This year we are asking everyone to use your computer's internal battery power for the survey. We are still providing the inverter – feel free to use it for your testing and any time up until the actual survey begins. This will make it easier to ensure that your battery is fully charged. Just flip the power switch on the detector off prior to starting the survey. You could unplug it if you like, but having the inverter there might prove handy if the battery runs low before you are done with the survey. The computer supplies power to both the bat detector and GPS. Neither consumes much energy. If your laptop's battery is in reasonably good shape and fully charged it should handle a full survey on its own. The great majority of computers have no problem using the inverter. If you are in mid-survey and get a warning that the battery is about to die, plug into the inverter, turn it on, and continue the survey. Everything will be fine.

Beginning the Survey
For a video demonstration, see StartToFinishSurvey.wmv, available in the video folder of the DVD. If you followed the installation and check out procedure exactly as written up to this point, the first time you re-start the DeLorme software you will see the displayed image of Salt Lake City that you saw when you installed the program. To recall your route (s):
1) Click the FILE OPEN icon (an image of a folder) in the upper left corner of the screen.
2) Click the map file you created when you loaded the routes. It will probably be named with one of your route names.
3) Click OPEN. You should now see the map as you saved it, and this is the way it will start up from now on. The route has both a start and finish point indicated, by green and red dots, respectively. You may decide to perform the survey in either direction. Forward and reverse direction routes are located on the DVD. Select the one that suits you best. See 9_GPSIntro1.wmv, available in the video folder of the DVD.

Whichever you decide, arrive at your start point well before the designated start time to allow yourself sufficient time to make sure everything is working properly.
1) Find a convenient parking place not far from the start location to install the detector and GPS antenna on the vehicle roof.
2) Turn on the laptop, using the cigarette lighter adapter for now if you like.
3) Connect the antenna and detector cables to the laptop and start the Delorme software. The map project will load automatically, with the same location and zoom level that you saved.
4) Click the GPS tab (lower left). Click **START GPS** and wait until the GPS indicator turns green or yellow, indicating that you have acquired a fix on your position.
5) If your map has the route activated (i.e., it shows on your map as a solid, highlighted line), the program will ask you if you want to navigate using the active route or hide it. Most people will have only one route on their map and can click on **NAVIGATE**. If you have loaded multiple routes, make sure the active route is the one you want to use that night.
6) At this point, if you have enough time before the planned start, you might wish to practice navigating the route, using the map to guide you.
       a. Begin driving the route so the driver and passenger can get used to the communication necessary to stay on the route. The navigator should become familiar with the information on the map display, including distance to next turn, current speed, etc. Make sure that you are back to the start point with plenty of time to begin the survey at the appropriate time!
7) After you are comfortable with route following, start the SPECT'R software.

42

8) Ensure that the device is properly connected. One good final check that everything is operating properly is to reach out an open window and jingle a set of keys. The blue-green display in the SPECTR window will jump and, if your computer has speakers and the volume control is turned up you will hear metallic sounds coming out of the speakers.
9) Click **File** (left side of the SPECT'R window) and enter a file name to save the data you will be recording. The default storage location is your My Documents folder. Create a new folder to store the files. Name the file [RouteName], where [RouteName] is your assigned route name. NOTE: There is no need to add the date to the file name because SPECTR automatically marks each bat call file with the date. If you are performing the survey more than once on the same route this year then add an incremental number indicating how many times the route has been surveyed, e.g., Albion2 indicates the second time you surveyed the Albion route this year.
10) Click **SAVE**.
11) You will be asked if you would like to start recording now. The easiest thing to do is not respond until you are ready to begin the route, at which point you will select **YES**. This is HIGHLY recommended. WARNING: If you select NO at this point, you *can* hit the RECORD button at a later time to begin the recording. Be aware, though, that it is VERY easy to get distracted with following the route, listening to bat calls, etc., and forget to hit RECORD. The display will show bat calls but nothing will be saved to the hard drive if you fail to hit the RECORD button. To avoid wasting the entire evening (not that anything like that has ever happened to me before) make doubly sure that you are recording at the start of the route. When you are in recording mode, the RECORD button is gray and inactive and the STOP button becomes active.
12) Turn off the inverter, if you haven't already done so.
13) After you begin the recording, and immediately before you begin driving the route, it is a good idea to run through the following, last minute checklist:
      a. Computer is powered from the internal battery, not its charger.
      b. GPS indicator is yellow or green.
      c. Device: AR125RevC
      d. Response: Natural
      e. Record Mode: Auto Snap
      f. File: [RouteName]
      g. Record Button: Grayed
      h. Stop Button: Active
      i. Test to make sure equipment is generally functioning: Reach out an open window and jingle a set of keys while holding them as high as you can reach. Watch bat detector display change while the keys are jingling.

Tips for performing the survey
Although it might seem that an automatic transmission would make it easier to maintain the desired 18-20 mph speed range, most people find that it's easier with a standard shift car. Either will work fine, though. In both cases, listening to the engine sound will likely prove helpful in addition to looking at the speedometer, but you will quickly determine what works best for you. Hilly sections are the most challenging when it comes to speed control. The passenger can read the speed on the GPS display and provide feedback to the driver if he slips out of range. To access this, click on the **GPS** tab. Unfortunately, you cannot simultaneously display both the turns list (Route tab) and the speed display (GPS tab). Some people have found that having a flashlight or headlamp is handy. This is mainly for the navigator, who might find it helpful for inspecting things without disturbing the driver by turning on the vehicle's interior lights. Similarly, most computers have the ability to dim the screen brightness and this is often helpful, both to extend battery life and to reduce the possibility of glare for the driver. You may wish to adjust the map zoom level to your personal preference. When you approach a turn while navigating with the GPS, though, the computer map will automatically zoom in to increase your ability to know where to turn. You will also hear a synthesized voice ("Microsoft Mary") warning you of the impending turn. You can quiet Mary's voice off by clicking the **Options** button at the top of the map screen, clicking the **Voice Settings** tab, and adjusting the **Voice Volume** to the desired level. How you arrange the map and SPECT'R windows will be a personal preference issue that also depends on your computer's screen size and resolution. The passenger can freely switch between the two applications. The best way to do this is via the Windows task bar, normally at the bottom of the screen. You can place the SPECTR window on top of the map, but on most computers you will have to obscure either the map or the turns list. Keep in mind that staying on the route is more important than watching the SPECTR display.
CAUTION: Be careful not to accidentally hit the STOP button on SPECTR while out on the route. This is easy to do, especially when changing the volume or display brightness. If this happens, the calls will still be displayed and you'll still hear them (if your volume setting is loud enough) but they won't be recorded. The safest thing to do is to minimize the SPECTR window, but if you want to watch the bat calls on the display feel free to do so. *Just be careful!* If you do accidentally hit the STOP button and notice it right away, just hit record and all will be fine. If, however, you notice that the recording was accidentally paused at some later time (or perhaps never started) you will have to start the survey over on another day. As you perform the survey, and assuming your computer's speakers are turned on, you will hear clicks, buzzes and bird-like chirps when you encounter bats. A graph of each call pulse is also drawn on the SPECT'R display. See Appendix 3 for an introduction to interpreting these.

Saving the data and shutting down
At the end of a successful survey you still have one more key step to perform: saving the data.
1) Click the **STOP** button in SPECTR. No further steps are necessary to store the files.
2) Click the **STOP GPS** button in Delorme Street Atlas. The "Save GPS Log File As" window will open. The default file name will be something like "NewLogFile2.gpl. Please change the name to **[RouteName]mm-dd-yy** and accept the default storage location, which is the GPSLogs folder. Click **SAVE**.
3) At some point, perhaps on the following day but certainly before you package up the data to send it to us, you should verify that the data were recorded properly.
      a. From My Computer, navigate to C:\DeLorme Docs\GPSLogs. You should see a file named [RouteName]mm-dd-yy.gpl. It will probably be about 150 to 350 KB in size.
      b. From My Documents, find the folder you created to store the bat call files. If you look in the folder you will likely see many files, all named similarly, something like [RouteName]_D20100528T0919m047.wav. Those are the recorded bat calls. Good job!

43

c. This is probably a good time to fill out the trip report, while everything is still fresh in your memory. Both the Driver and the navigator need to fill out the report. THIS IS IMPORTANT. Please ensure that both names and signatures are on the report, and that all of your valuable time is properly accounted for.

Frequently Asked Questions

1) What if I arrive late to the start of the route?
a. Missing the start time by a few minutes is no big deal. If you are more then 10 minutes late, though, you must perform the survey on another day.

2) Can I reverse the route, starting at the planned finish location?
a. Yes. The route library contains two versions of each route, one for each direction. One is named [RouteName].anr, the other is [RouteName] reverse.anr. Simply choose the one you prefer.

3) What do I do if I miss a turn?
a. Unless you end up driving for 10 minutes before you realize the error, which is pretty unlikely if you are regularly checking the map, missing a turn or two (or three) is no problem. Simply turn around as quickly as it is safe to do so and get back on the route, preferably at the point where you made the error. If that's not possible for some reason, don't worry about it. Note the occurrence in the trip report and describe what you did to get back on track.

4) What if it starts to rain while we're out on the road?
a. You must terminate the survey. Pull over as soon as it is safe to do so, remove the bat detector from the roof. Save the files exactly as if you were able to complete the route. Note the occurrence on the trip report. Plan to resume the trip from the beginning at a later date.

5) How close to the desired 18-20 mph do I have to stay?
a. Experience suggests that it's pretty easy to stay quite close to the desired range so long as the driver can count on the passenger to navigate. If you have to err, slower is better than faster and dropping to 15 mph or so for brief periods every once in a while is no problem. In short, do your best and don't worry about it.

6) How fast can I drive with the bat detector mounted to the roof?
a. Although we need to be careful during the survey not to exceed the 20 mph upper limit except occasionally and then not by much, there may be times (before or after the survey, for example), when you would like to drive faster. It is best to avoid highway speeds (60 mph or higher), but the magnet mount has been designed to withstand any reasonable speed you are likely to drive.

7) What if it is necessary to take a long stop during the survey?
a. Short stops are not a problem, although it would be best if you describe any stop longer than a minute or so in the trip report. If the total time stopped exceeds 10 minutes then it may be necessary to repeat the route at a later date. Contact us and we can discuss it.

8) What if there are portions of the prescribed route that could not be surveyed (road closures, etc.)?
a. You'll need to use some judgment here. Brief deviations that still allow you to cover 75% or more of the route as planned are no great problem. Simply note them on the trip report. In most cases, when departing from the planned route the best thing for you to do is to drive at whatever speeds you normally would (i.e., if you were not doing a bat survey) until you are able to return to the planned route. You should not, under any circumstances, attempt to drive at the speeds prescribed by this study on divided highways, roads with a posted minimum speed limit, or any road on which it is unsafe to do so. Remember, safety comes first.

9) What if the GPS receiver loses a position fix at some point during the survey?
a. Occasional dropouts (when the GPS indicator turns red and stops drawing your path on the map) can be expected, especially on the tree lined roads that were favored during the route generation process. This is not cause for concern, although you may eventually have to manually advance the turn list to know where to make the next turn. If the GPS indicator is red for more than a minute then something is probably wrong. Check to see if the antenna is properly installed.

10) What if I accidentally hit the STOP button in the SPECTR window during the survey?
a. If you notice when it happens and immediately hit the record button again then there is no problem. If, however, you discover that it is not recording and aren't sure when it happened then you will have to re-start the route from the beginning, which will usually have to be done on another day.

11) What if I start a survey but am unable to finish for some reason?
a. In general we would like you to attempt to survey the complete route at a later date. If this is not possible, please submit the data as if you completed the route but note the circumstances of your deviation from the planned route on the trip report.

12) What are those clicking, buzzing and chirping sounds I hear when doing the survey?
a. Those are bat calls! You can adjust the volume on your computer (or even silence them) if you wish, but most people like to listen to them.

13) Can I listen to the bat calls but silence the computer voice that warns about the upcoming turns?
a. Yes. To quiet "Microsoft Mary", choose the OPTIONS menu in the DeLorme Street Atlas program (found near the top of the screen, next to the yellow question mark button). Click the VOICE SETTINGS tab and find the VOICE VOLUME control, normally set at 100. Change the setting to 0, or as low as you like.

Appendix 1 - Default Settings for SPECT'R

The disk provided has a customized version of SPECTR with default settings of all of the recording and data storage parameters set to the values used for this study. **Under normal circumstances, therefore, it should not be necessary to adjust any of them.**
For reference in troubleshooting, here is a step-by-step list of the correct value of those settings:
Setup Procedure:
1) Setup/Time standard: Local
2) Setup/Device: USB 0
3) Setup/Response: Natural
4) Setup/Record: Auto Snap
      a. Snapshot Trigger Threshold: Automatic
            i. NF+ 15 dB
      b. Trigger Range

       i. Low: 15 kHz
       ii. High: 120 kHz
    c. Snapshot Duration: 20 seconds
    d. Idle timeout: checked, 1.0
    e. Pop Filter: On
    f. Output Format: Standard
    g. Time Expansion Factor: 1
    h. Comment: <no comment>
      i. Close Record Setup Window
5) Window/SonoSCOPE
    a. Button should say "Pause"
    b. Threshold: slide all the way left
    c. Detail: Med
    d. Divisions: 4
    e. Marker 1: checked, 40.0
    f. Marker 2: checked, 25.0
    g. Close SonoSCOPE window
6) Window/Spectral Trace
    a. Adjust green slider (right-hand side of main Spect'r Window) so center mark is at 30 kHz
    b. Peak Hold: unchecked
7) Volume: [personal preference]
8) Brightness: mid-range, approx. 50
9) Contrast: mid-range, approx 50

Appendix 2 – Dealing with minimal available hard drive space
Between the mapping software and the amount of space needed to store the bat calls, this project is fairly intensive in its use of hard drive space. If you are short of space and don't know what to do about it we have some options that may help.
1) Don't install the maps to the hard drive. When installing the DeLorme software, you have the option of whether you want to store map data to the hard drive or not. Putting the maps on the hard drive is the most fool proof method, but it is possible to run the DeLorme software without doing this if you keep the disk in the disk drive while your are doing the survey. This will reduce the amount of hard drive space necessary by more than 1.5 GB.
    a. In addition to making the 2 standard changes when the Custom Setup Window appears (those being activating advanced file management and turning off the automatic program updates), click the box next to **Install Street Atlas 2009 Road Data.**
    b. Select the option to not store the road data and map features on the hard drive. You will need to have the disk in your drive to use the map.
    c. Click Next
2) Store the bat calls to another drive. Again, storing the bat calls to the hard drive is the recommended method because there is less chance that something will go wrong, but if you have a third USB port available you could plug in the thumb drive that we provided to submit the data and store the bat calls directly there. To do this, have the drive in place when you start the SPECTR software. Follow the normal setup procedure described above. When you get to the step where you click FILE and enter the route name, navigate away from the default "MY Documents" folder and find the thumb drive. It may appear as your computer's F: drive, but this varies from one computer to another. Be sure to name the file as described in the previous instructions, [RouteName]-mm-dd-yy.

Appendix 3 Making Sense Out of Bat Calls
Although there is no need for you to understand any of this in order to participate in the study, many people are curious about the funny graphs and sounds they see and hear during the survey. If you are one of those people, read on. Most folks know that bats use SONAR to find their way around in the dark, and that the sounds they emit are generally too high-pitched for human ears to detect. The bat detector we are using acts like a microphone, allowing us to record these calls. The software also displays the calls visually as it is recording them. It also attempts to translate the calls down to a frequency range that we can hear. Among other things, this provides useful feedback that things are working properly.
1) The primary display of the SPECTR software is a graph that plots the frequency of echolocation pulses (vertical axis) versus time (horizontal axis). The default display configuration breaks the graph into 4 panels, each 25 milliseconds wide. The vertical scale on the right side of the graph indicates frequency, in kilohertz. For reference, most people can't hear sounds above 15 kilohertz.
2) The software tries to plot one echolocation pulse per panel.
3) Varying colors on the graph indicate loudness. Blue is relatively quiet, green is louder, yellow is louder still, and red is the loudest.
4) If your computer has speakers you can listen to some of the bat calls. SPECTR takes a portion of the frequency range that bats emit and translates it down so that humans can hear it. The presentation is not particularly useful or engaging, but it does give the driver some idea of how often you are encountering bats without having to look at the computer screen.
5) how do we distinguish between individuals?
    a. Most encounters are brief because the effective range of the detector is rather limited. Bats soon fly out of range.
    b. Sometimes a bat will be flying in the same direction that the car is traveling, in which case the recording might last 15 seconds or more before the bat veers off the route, but that's about it.
    c. We are driving at the approximate top speed for most of these species. This coupled with the fact that the routes are laid out in a generally straight line means that after a gap when no calls are encountered, every new encounter is another individual.
6) how do we distinguish between species?
    a. Bird song and other calls are commonly thought to serve as forms of communication. Although bats also sometimes emit calls that seem to have a somewhat similar purpose, those that we are interested in are echolocation calls, used by the bats

to help them fly and feed in the dark. In particular, the calls that tell us the most about which species is responsible for generating them are known as search-phase calls, emitted when bats are flying about in search of insect prey.

b. Two factors that heavily influence a bat's search-phase call characteristics are the size range of the bat's acceptable prey and the amount of clutter (branches, tree trunks, etc.) in its typical foraging area.

  i. Big bats tend to favor big prey and it's more efficient for them to search for big prey with lower pitched calls. Our largest bat, the hoary, emits calls that are so low in pitch that some (young) people with exceptionally good hearing can detect them with the naked ear.

  ii. Bats that tend to forage in open areas usually favor a call with relatively constant frequency. Bats in cluttered areas tend to emit calls that start out at high frequencies but that slew sharply downward in frequency. Some species tend to "flatten out" (see figure 4, for example) toward the end of the call and others do not.

  iii. Some bats are generalists when it comes to their feeding preferences and emit calls that vary depending on environment they happen to be using at any given time. The preferred prey of the day probably influences this as well. Other species are specialists and these have less variability in the nature of their calls. Some bats, those that heavily favor forested areas, for example, emit clutter - calls even when they are flying in the open.

7) insight into bat behavior

a. When operating in search-phase, bats typically emit around 10 sound pulses every second. The pulses are quite short in duration, about 10 to 20 milliseconds or so. If you do the math you'll realize that means that most of the time they are silent, listening for echoes to come back from their last pulse. The faster the echo returns, the closer the bat is to the item that the echo bounced off of.

b. Bats can catch insects on the fly in total darkness. When they detect an interesting sounding echo they fly toward the source. As they get closer, the echoes come back sooner. The bats then send out shorter and shorter pulses more and more frequently, which gives them increasingly precise feedback on where their intended prey is located. Just prior to capturing the insect/meal, the pulses are very brief and very rapid: approaching 200 pulses per second in some cases. This is referred to as a feeding buzz, because of how it sounds when heard with a bat detector that has audio output capability.

c. Bats' high-energy lifestyle means they need to feed often, consuming a significant percentage of their bodyweight each night. Feeding buzzes are, thus, quite common.

Appendix 4 – Contents of the Survey Kit

Survey Kit contents:
Roof top magnet mounted bat detector
USB cable to connect bat detector to the computer
Software disk
Includes software for the bat detector, GPS, route descriptions, operating manuals, electronic versions of forms, training videos, an electronic copy of this document.
GPS antenna
12 VDC to 110 VAC inverter
Thumb drive, to hold the data for submission to the project coordinators
Trip report forms

I want morebooks!

Buy your books fast and straightforward online - at one of the world's fastest growing online book stores! Environmentally sound due to Print-on-Demand technologies.

Buy your books online at

www.get-morebooks.com

Kaufen Sie Ihre Bücher schnell und unkompliziert online – auf einer der am schnellsten wachsenden Buchhandelsplattformen weltweit!
Dank Print-On-Demand umwelt- und ressourcenschonend produziert.

Bücher schneller online kaufen

www.morebooks.de

OmniScriptum Marketing DEU GmbH
Heinrich-Böcking-Str. 6-8
D - 66121 Saarbrücken
Telefax: +49 681 93 81 567-9

info@omniscriptum.com
www.omniscriptum.com